Six Sigma Series

Getting Ready for Six Sigma

Lean

Klaus Hogreve

DEDICATION

This print version of my books is dedicated to the few old fashioned people who still prefer a print version over a digital download.

CONTENTS

BOOK ONE:
GETTING READY FOR SIX SIGMA

BOOK TWO: LEAN

ACKNOWLEDGMENTS

Special thanks to Jürgen Blankenburg for providing me with a vast amount of information, which enabled me to write this books, and to Kamran Avary, PhD for inspiring me to write the Six Sigma Series.

Book One

Getting Ready for Six Sigma

PREFACE

Six Sigma is a rigorous process comprised of DMAIC (Define, Measure, Analyze, Improve, Control), designed to reduce the errors (mistakes) to a rate of 3.4 per million opportunities or better. It's the war against variation. The main tool of Six Sigma is statistics. It seems that there are more books written on Six Sigma and Lean Six Sigma than there are stars in the universe, so I won't elaborate on this any further. If you want to read about Six Sigma, I recommend Thomas Pyzdek: The Six Sigma Handbook, or for the ones who just want a short overview Jay Arthur: Lean Six Sigma DeMystified. For an even easier entry into the subject there is Barbara Wheat/Chuck Mills/Mike Carell: Leaning Into Six Sigma.

The idea to this book came a few years ago from talking to a friend who is like me a Six Sigma Black Belt. He just started a new assignment and told me that after a few months his accomplishment was to implement 5S and that he wasn't even close to run a Six Sigma project. This goes confirm with my own experience, getting assignments offered to be part of a team of Black Belts to do process mapping.
What a wrong allocation of recourses! Black Belts should run Six Sigma projects, not do the groundwork. Given the professional rates in the industry, the budget will be blown on basic tasks. This kind of tasks would be better handed to an engineering intern or a new hire fresh from School to get him/her used to the company.

Therefore the emphasis of this book is to get the organization ready for a Six Sigma project; to do the easy work upfront, so the specialists can start with a real Six Sigma project and don't have to spend their valuable time with preparation.

I will also give some guidance for assessment of the organizations readiness for Six Sigma. Realistically, it can take years or even decades to get an organization ready for Six Sigma. Implementing Six Sigma Projects prematurely has its risks, but waiting too long can be even more disastrous.

PROCESS MAPPING

Every business has its processes. That is what makes it a business and allows it to be efficient. Otherwise it would be "garage firm" just starting out and figuring out how to do things. Established processes make it an established business.

These processes are documented in job descriptions and on a more global basis thru process mapping; a flow chart. This involved defining what the business is doing, who is responsible for what, and to what standard a business process should be completed. This will be done starting from a global view down into the details of the sub processes. It also will show how the processes are interconnected. How detailed it will be documented depends on the business, the individual process, and willingness to detail work. The process map will enable management and outside consultants to determine the effectiveness (the right process is followed the first time), efficiency (continually improved to ensure processes use the smallest amount of resources) and will show where a need or room for improvements is.

Therefore the process map is an essential base for a Six Sigma project.

The four major steps of process mapping are:
1. Process identification - attaining a full understanding of all the steps of a process.
2. Information gathering - identifying objectives, risks, and key controls in a process.
3. Interviewing and mapping - understanding the point of view of individuals involved in the process and designing actual maps.
4. Analysis - utilizing tools and approaches to make the process run more effectively and efficiently.

Below an easy example of a process mapping for making breakfast. Please note that the details are only outlined for one activity at each level. Each activity should be mapped in detail.

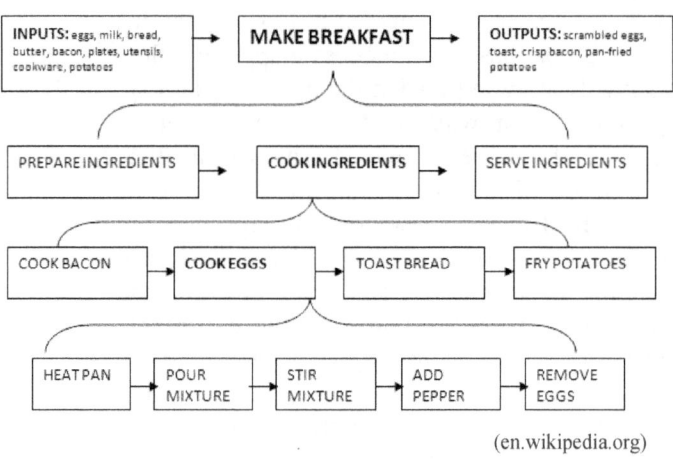

(en.wikipedia.org)

The most commonly used symbols are
- start / end box
- activity box
- decision diamond
- connecting arrow

which are standardized under the ANSI standard.

The following is an example of a simple global cross functional process map. It is cross functional because of the extra dimension of the departments involved.

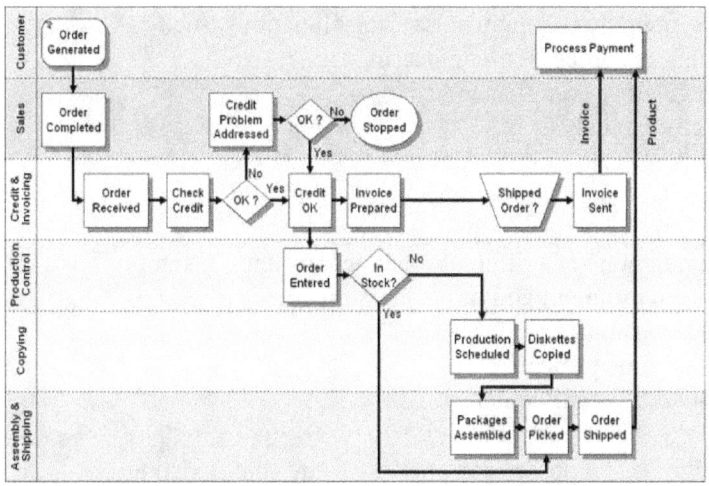

(from mybusinessprocess.net)

For each activity box there should be another more detailed process map created. In reality a detailed process mapping system will be very complex. Most of the time a flow chart diagram, like the one above, is used.

A flow chart shows the

7

- inputs (resources such as land, materials, labor, equipment)
- actions (procedures, processing, handling, transportation, storage, and processing)
- outputs (products, services, rework, scrap, pollution) in a graphically way.

Another dimension which could be added is a time marker, either as a time graph following the process or as process duration shown in the upper corner of the activity box.

Process mapping is supporting various functions, such as
    -Business process re-engineering
    -Regulatory compliance
    -Activity analysis
    -SLA (Service Level Agreement – the essential part of a service contract with formally defined service expectations)
    -Simulation of different changes in the business environment and/or business processes
    -Internal audit
    -Six Sigma projects
    -ISO9100 certification

The process mapping can be done by the existing personnel and process owners, but because of the time consuming nature of the project, it may be a good project for interns or to get new hires accustomed to the organization.

If the organization has already its process mapping done, it should be reevaluated from time to time to make sure it still reflects reality, since everything is evolving. For the organizations "blessed" with the SOX requirements, a regular reevaluation is already a requirement. In the presents of an upcoming Six Sigma project a re-evaluation will be also good practice.

VALUE STREAM MAPPING

Value Stream Mapping is similar to the Process Mapping above; just the view has now changed to the perspective of the customer (internal or external customer). Is CAV (Customer Added Value) produced? BNVA (Business Non-Value Added) or NVA (Non-Value Added)? The customer defines value, not the department creating it. Since most businesses have grouped work into functional silos, the temptation is great that every department defines the created value as they please. This is wrong! Therefore the focus is on how the customer values the output.

The role of the value stream mapping is
- define value from the customer's view
- map the current state of the value stream
- apply the tools of Lean in order to eliminate muda (waist)
- map the future (desired) state process
- develop the transition plan
- implement the transition plan
- validate the new process

To map the value stream

1. Start by indentifying customer needs and end with satisfying them.

2. Use square Post-it notes to layout the processes

3. Use arrow Post-it notes to show delays

4. Place activities in the correct order

5. Identify inventory levels carried between each step.

Depending on the complexity of the operation, the value stream will be mapped for each process and sub process..

Current-State Value-Stream Map

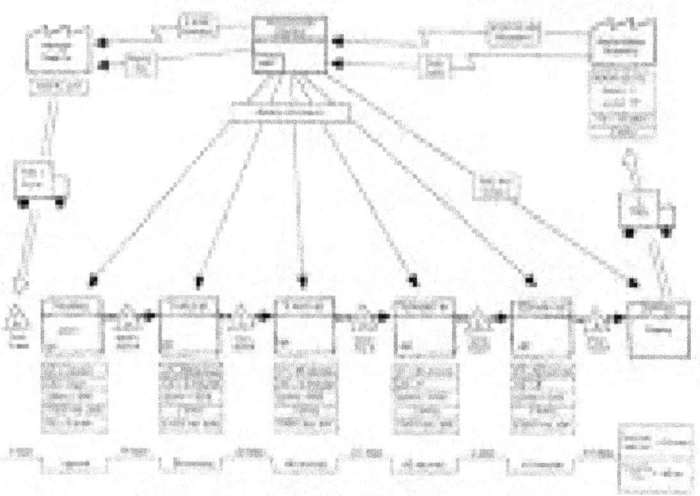

Future-State Value-Stream Map

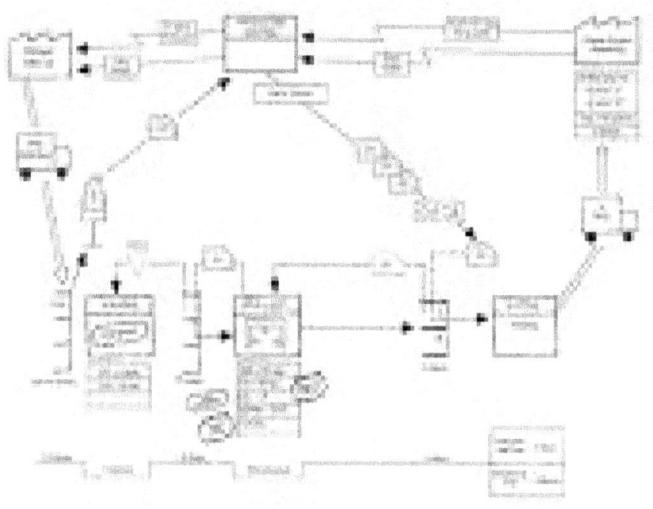

As shown above, the elements of a complete value stream map are:
- Process diagram
- Material flow
- Information flow
- Process data
- Process Lead Time and Value Add Time

The value stream mapping and the attempt to improve on it will result significant improvements. It is a long term strategy commitment. This is contraire to the improvements made bases on process mapping, which are usually small and based on short term / tactical planning.

Value Stream for a Project with Remote Testers and Customers: How long does it take to finish a feature?

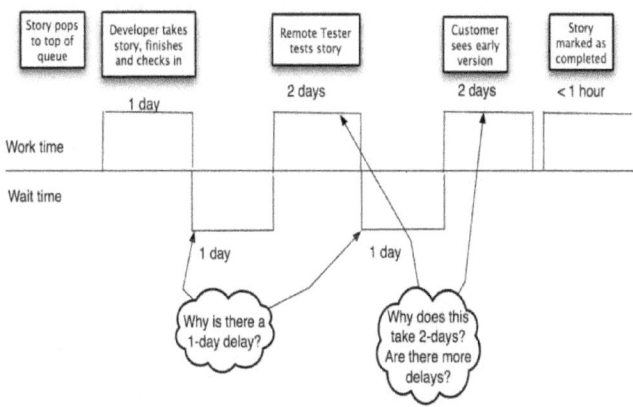

www.breezetree.com

One byproduct of value stream mapping as well as process mapping is that it has the potential to show how much time a product is sitting around waiting for the next step. This also becomes obvious when one follows the product from start to finish. Often this is an eye opener.

From the process mapping and value stream mapping it's just a small step to a **Value Added Flow Analysis**. Across the top row the persons or department involved will be listed. Along the left hand side the major steps in the process (planning, doing, checking, and acting to improve) are shown. Starting from flow charting the process, every action, decision and arrow in the flow chart will be evaluated to identify the non-value added activities.

DATA COLLECTION

The key of a Six Sigma project is data. As mentioned before, Six Sigma uses statistical tools to improve the processes. It's all about variation, which shall be minimized. In order to use statistics, one needs data. Therefore the data is essential for the project. Some relevant data may be already collected, some is not. Every process needs to get measured. It's the old saying "If it isn't measured, it won't happen". Measuring shows the associates that management cares about it. If the results measured are publicly displayed everybody knows it's important and that management cares about it. It is important that those reports are distributed to the managers for that specific area.

I t would be nice to have all the data needed for the Six Sigma project ready to hand off to the consultants, but it is hard to predict what they are going to need. But something available is better than nothing.

There are three different kinds of data
1. Internal data – to be collected by measurement and recordkeeping
2. External data – to be collected by surveys (like VOC (Voice Of the Customer)
3. Public data – readily available on the internet

Internal data

What and how to measure depends on the individual situation. It should be the key data management cares about. In most cases the output is getting measured. For the purpose of Six Sigma not only the output needs to get documented, it also needs to document the defects. From a Six Sigma standpoint all what matters is the value "good" or "defect", not the extent or form of the defect. When there are plenty of defects, the measurement should be extended to the sub processes. In other cases the cycle time (duration a process takes to complete) will get measured, for example in a call center, and in other situations the input and output will get measured, like for a motor vehicle the input (fuel) and the output (distance driven). It all depends on the process and the situation and should include the relevant areas of variation. The data collection should have some meaning to the process and the organization.

As stated before, one goal of the data collection is to have something to hand over to the Six Sigma Black Belts to start with; another goal is to create awareness and acceptance in the organization of basing decisions on data.

In most pre - Six Sigma organizations decisions are often based on assumption, gut feelings or the famous "we doing it this way for over 25 years". In Six Sigma organizations the tonus is "show me the data". Decisions are based on hard data. No decision will be done without the data backup. This way of thought will have to be "handed down" from top management all the way to every team member. Actually, management has to practice it! This is an important cultural shift in the whole organization which takes some time to adapt to.

Another problem will be that as soon the modus operandum is shifting and decisions are based on data, there will be the notorious "naysayer" stating that the data is not correct. This usually will happen in meetings when data is presented in order to support decisions. Specially if the data makes her/him look bad or the logical decision based on this data is not the one he/she is favoring. But since the data is already collected for some periods and was distributed to the managers it should concern (just another report), this person is easy to silence by pointing out that this is no new data and he/she didn't object this data in the past.

External data

The main source of external data should be the voice of the (external) customer (VOB). This can be gathered by surveys, customer interviews, or reports from the sales organization.

There are two challenges with data collection and statistics: Biased data collection and creating the wrong correlation. Both result in what I call "lying with statistics".

Bias happens when the data collection is tampered with. I remember back in Business School when a friend

17

was running a marketing project and called me and some other friends to come by to fill out a survey because he needed some specific answers for his analysis. In order to be reliable the data cannot contain bias and it is important to collect all data to the same standard.

When I was growing up in Europe an April 1st study was published proving a strong correlation between the populations of stocks and the regional birthrate. From a statistical standpoint this study was flawless, it just was wrong to correlate two not related factors. Therefore it is good practice to make sure the factors really influence each other in real live before correlating them in a statistical model.

Public data

Public data is available from multiple sources, usually accessible within minutes and should not be the concern at this time.

GOING AFTER THE LOW HANGING FRUITS

There is a large improvement potential by just using common sense and a few simple tools. This is called going after the low hanging fruits. There are many tools available but I will only introduce a few at this point, as I focus on the ones that are relatively easy to apply and have the potential of great returns.

Constraint Management - Bottlenecks

It is always worth to look into bottlenecks, but if it seems to be too complex to ease bottleneck restrictions, it may be not worth it to deal with right now. I recommend to do easy project now (low hanging fruits) and leave the other ones to the Black Belts.

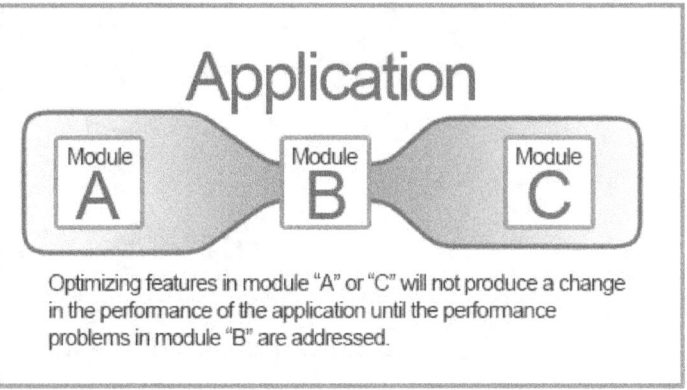

Optimizing features in module "A" or "C" will not produce a change in the performance of the application until the performance problems in module "B" are addressed.

A potential bottleneck is easily identified in the process or value stream mapping as well as just going on the floor and looking were inventory piles up or were insufficient supply slows down the process.

But due diligence requires to verify that it is a real bottleneck before taking action.

There may be a bottleneck shown on the paper in the process or value stream mapping, but if the facility never or rarely reaches the capacity of the bottleneck, it makes no sense acting on it at the present time.

If production slows down due to leak of supply or inventory piles up at one station, there may be another reason for it too. A short seasonal peak or the unplanned downtime of a machine may be responsible.

It is good practice to find out the root cause of the problem before acting to solve it. We want to cure the cause, not the symptom.

5S

The most potential is usually unfold by using the 5S method (Sort, Set in order, Shine, Standardized cleanup, Sustain).

Set in order
- Clean up all unnecessary things.
- Don't store anything on the ground.
- Keep walkways free of stuff and debris.
- Don't put things against walls, machines, e.g.
- Keep the work surface free of clutter.

Sort
- Keep frequently needed items in arms reach.
- Have a designated place for each tool and material (mark that space accordingly (shadow board) so one can see if it is missing) – this prevents the need of digging for a long time in the tool chest in order to find a specific tool.

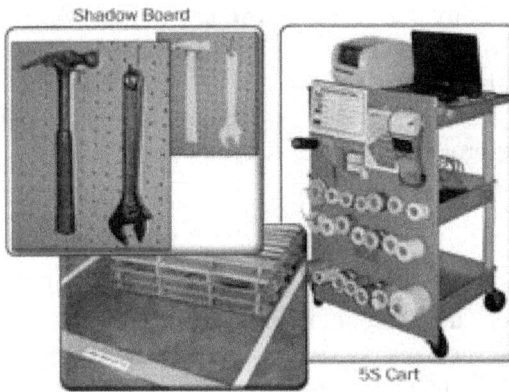

Shadow Board

5S Cart

Floor Marking

www.duralabel.com

- Keep an inventory of drawers, tool chests, storage spaces, e.g.
- Store shared tools and supplies in a central space,

It is important that every item is retrievable in a minute or less.

Shine

- Clean up dirt, debris, and trash around the work space and machines.
- Keep all tools and machines clean and in good working order.

Cleaning enables the detection of weaknesses on machines in the working environment in an early stage. The benefits are the avoidance of

- accidents
- unexpected breakdowns
- hidden expenses due to inefficient operation of machines (higher consumption of energy or other recourses)
- lower quality due to higher variance

because repair or maintenance can be scheduled before serious consequences arise. For example the early detection of a leak of a bearing or seal in a hydraulic system and appropriate action (repair request) will prevent a breakdown or accident.

Standardize

- Set up of cleaning and maintenance schedules and keeps documentation of it visible.
- Identify the persons responsible for each task / area. Each team member should be responsible for his / her area.
- Define, mark, and label all storage spaces on the work desk and the work area.

- Define min and max inventory and keep this information visible.
- Create a layout in order to mark areas of responsibility.

Sustain
- Define the expected standard and make it visible. Create SOPs (Standard Operating Procedures).
- Continuously improve the standard.
- Make every team member responsible for compliance with the standard.
- Provide training, verify compliance, and enforce discipline.

Going thru this exercise for every work station / work space will ensure that all are well organized and ready to focus on the task, not wasting time on getting ready for basic tasks and trying to find the items needed.

Lean Design

Under lean design I understand the design of a process using simple methods to help speed up the process or prevent errors.

Color coding

One element of lean design is color coding. It's already all over the place. Why not using it in this situation too?

Electrical cables are color coded for a reason: blue is "hot", black is the return, and yellow/green is "ground". Red, yellow and green lights have a universal meaning. On your faucets there is a blue symbol for cold and a red symbol for hot water. Everybody knows this, is there is no surprise. It's a standard even outsiders are aware off. The same can be implemented in every other area. My pharmacy puts color coded tags on the medicine bottles, a different one for every household member so they are easily identified by for whom they are. Laboratories still use color codes to identify the process each sample has to go thru (even they now rely more on barcodes). Grease nipples can be color coded to show to which maintenance interval they belong to. Handbooks are color coded to identify different content.

Checklists

Checklists can be a great improvement tool too. Sometimes checklists are very complex, which may have its legitimacy, but in this context I mean short checklists with up to 5 items following a product or process. Their propose is just to show the advancement in the process so that one can see within a second what the status is, eliminating the need to read thru some tags or lists. Color coded tags will do the same, like triage tags in the disaster response situation.

Follow the product

As pointed out before, physically following the product is a great tool to identify waste. In this case waste of unnecessary movements and the waste of wait time.
Just following a product, an actual product or a form to be administered or approved from start to finish through the organization can be a real eye-opener. Documenting each step with actual times is a great way to update the process mapping.

If it is not practical to physically follow the product, like an email or online approval process, it is still a good idea to follow the chain of stations it will go thru and document it, as well as the wait time on each.

Ergonomics

Ergonomics seems to be the step child of process improvement. It is hardly mentioned. There are two major areas of ergonomics:

- Workplace ergonomics
- Process flow ergonomics

Workplace ergonomics

The classical workplace ergonomics deals with the ideal height of the chair in relation to the desk or workstation, the ideal placement of the computer monitor and keyboard e.g.. There is plenty of books and article written about this. OSHA in the US and the major health insurance companies in Europe usually provide free information, sometimes even on the job assessments and evaluations or online support.

But there is more to it: 'Soft factors" such as the right lighting and something I call workplace organization. It's part of the 5S element "sort". It is the organization of the workplace in a way that unnecessary tasks or sub-tasks are eliminated. Everything needed should be placed within easy reach. This avoids the unnecessary looking, searching, sorting what can add up in time wasted as well as the physical movement associated with it.

Process flow ergonomics

Almost every process has preceding and downstream processes. They all should be in harmony, going hand-in-hand, without delays and long ways to move from one step to the next. They should – but do they?

I remember from my hospital visits when I was young that the ER and Radiology were on the opposite site of the building, preferably even on separate floors. Perfect, since most accident victims will have to get an X-Ray. That was in the old days. Luckily they now improved even further than expected and have the Radiology on a little card going from patient to patient. A quantum leap improvement; now the equipment comes to the patient instead of the patient being hauled across the building, waiting in another line again.

The floor plan of the modern American house has a process flow / lean perspective too. The garage is next to the kitchen, ideal to move groceries from the car to the refrigerator in an efficient way. Ironically, most families use their garage for everything else but to park the car. But the thought is what counts.

Similar to the workplace ergonomics, process flow ergonomics avoids extra efforts by designing the processes and the lineup of their stations accordingly.

Can we learn from the above examples and redesign our processes for a better process flow?

ASSESSING THE ORGANIZATIONS READINESS FOR SIA SIGMA

According to ISixSigma, there are two main areas of readiness of organization for Six Sigma - cultural readiness and organizational readiness.

But mainly and most important: Does the organization have stable processes? Without a stable process there is no point of starting a Six Sigma project! This question should be answered even before considering a Six Sigma project. Without a stable process there is no way to do a meaningful statistical analysis because the data is too random. If a stable process is not the place, it is strongly advised to fix this problem first. Process mapping as well as policy and procedure efforts will help with this problem.

Cultural Assessment

The goal of the cultural assessment is to identify whether change will be accepted across the organization. This assessment can vary employ a range of tools from simple observation to a formal survey of all employees and it should answer the following questions:

- How do senior leaders communicate important information throughout the organization?
- How are important decisions made?
- Who makes the decisions?
- To which grade are the decisions fact (data) based or based on assumptions?
- How fast are the decisions implemented?
- How does the organization recognize successes and failures?
- How does the organization handle failures?
- Does everyone in the organization understand the mission, vision and strategy of the firm as a whole?
- Is everyone in the organization aware of the critical customers, revenue and operating expense issues?
- How does the firm set up corporate goals?
- How clear are these corporate goals?
- Are the corporate goals measurable?

With most of this questions answered:

How is the current state of the organization supporting the changes coming with a Six Sigma Project?

Will a Six Sigma project be supported or boycotted?

Operational Assessment

An operational assessment measures the maturity of the processes, measurement systems and data systems in the organization and must answered the following questions:

- How does the organization measure success?
- Does the organization measure the right things?
- How often does the organization measure these things?
- Does the organization have a few metrics that all employees understand and use?
- Are decisions based on data or assumptions?
- Who owns each critical process?
- Is data stored on spreadsheets on employee laptops or in a data warehouse?
- Has the data been validated?
- Are the company's reports written in simple, scientific or financial terms, or free form using fuzzy language?
- Are there updated process maps of the most critical processes?
- Do the executives know what a process map is and understand it?

With most of this questions answered:
Does the organization have the support system needed in place? This includes the key players mentioned in the next chapter, financial resources, and the support of the data owners.
Is the organization able to provide the data support and the reliable data needed for a Six Sigma project?
Are the data owners supportive or not?

But even more importantly: In case the organization is not ready yet to take on a Six Sigma project, this analysis will provide the organization with the knowledge needed to build a tailored deployment plan to get ready for Six Sigma. The deficiencies are identified and the improvement plan should include the right elements needed and mitigate the risk of failure.

IDENTIFYING THE KEY PLAYERS OF THE SIX SIGMA PROJECT

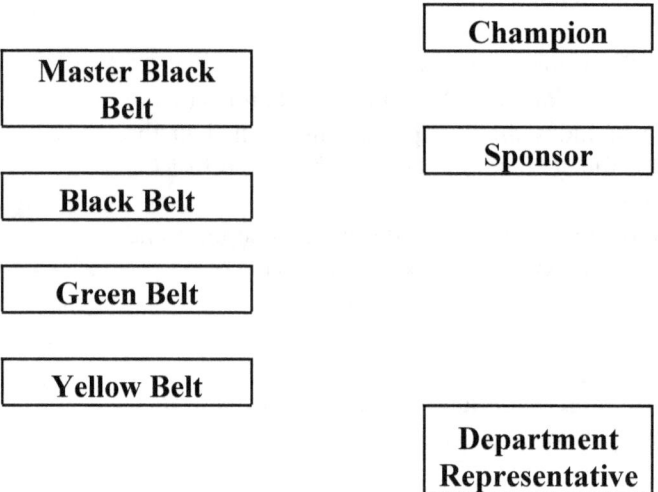

Each Six Sigma project needs people to run it. Some will initially come from the Consulting firm (external staff), that are most of the ones on the left hand side, but the one on the left hand side will have to come from the organization which is running the project.

As the organization gains experience with Six Sigma projects, it will also gain the internal expertise and some members of the team will get Certified and become "Belts". Ultimately, the organization will grow their own staff needed to run Six Sigma projects and the Consultants will phase out.

External staff

Master Black Belt – will be from the consulting firm, has a deep expertise of mathematics and statistics, coaches the Black Belts and communicates which to management.

Black Belt - will be from the consulting firm, runs the project and has a deep expertise of statistics.

Green Belt - will most likely be from the consulting firm or in house trained, assists the Black Belt and is capable to form and facilitate project teams and manage projects.

Yellow Belt – most likely will be a trained staff member from the organization assisting with the data collection which has limited statistics expertise.

Internal staff

Champion – top manager of the organization supporting the Six Sigma philosophy and projects. The Champion is often involved in the selection of the projects.

Sponsor – the manager of the organization who is supplying the projects and the resources for the project. The sponsor is also participating in all project reviews.

Department Representative – staff of the organization involved in the project. Most likely this person is a key employee with a deep knowledge of the processes the project is about. Some Black Belts select also department staff from other departments in order to get a different view of the problem and from accounting to verify financial data.

To prepare for a Six Sigma project, it is a good idea to identify this key player on the organizations side. The champion may be already identified by the fact that a Six Sigma project is approved, but there is still the sponsor to find, the one who takes the burden of the project on the management side. Identifying the department representatives will be a little more time consuming task:
- Specify the roles
- Identify the persons
- Verify with these person's the interest to join the Six Sigma team
- Get permission from the supervisors of the prospective team members to join the team

Having this legwork done upfront will speed up the process and save valuable time when the consultants are there.

IDENTIFYING PROSPECTIVE PROJECTS

Identifying the right Six Sigma project is not as easy as it seems. Sometimes management has already selected the first project(s), most likely the reason for starting Six Sigma. But sometimes this is not the case. And at least after the managements request for specific projects is satisfied, the Six Sigma team has to find them on their own.

A valid potential project can be every project that results in a process output having an error rate higher than 3.4 per million. 3.4 errors per million opportunities is the target of Six Sigma. But in reality there are hundreds or even thousands of possible projects in the organization. Six Sigma is not limited to the production area, it can (and should) be used in any area. The only thing what's needed is a stable process and measurable outcomes.

Because of the multitude of potential projects the projects have to be ranked. Usually the ranking will be done by weighted factors. Some of the most used factors are:
- Level of difficulty of the project
- Number of people needed to run the project
- Prospective duration of the project
- Possible payout (savings) of the project

In the beginning it is important to select projects which promise an easy win. They shouldn't take long and shouldn't be too difficult. But the savings the project produces should be worth the effort.

It is important for the whole Six Sigma program that the team can show early success in order for the program to be accepted as a valid tool and the program to succeed. If the project takes too long and doesn't show success within reasonable time, management will withdraw support, the team members get demotivated, and the naysayers will say "I told you so".

EXAMPLE OF A SIMPLE SIX SIGMA PROJECT

The following example is to illustrate the steps of a Six Sigma project and the bare minimum of what is needed. This is a really simple and stripped down project. A real project will most likely be way more complex.

Case: A manufacturing company is fabricating metal cans and containers on a large scale. The VP of sales is complaining that his sales staff is getting a lot of calls from customers regarding incorrect billing. This leads to extra work (lost productivity) and stained customer relations.

Management decides that this should be solved as a Six Sigma project. For this reason there is no need to evaluate other potential project to seek the one with the highest possible return. It is estimated that each wrong invoice costs about $100.00 to correct, from taking the call from the customer to mail out the correction (hard costs.) This does not include the price of lost good-will from the customer, even so the good-will factor is considered more valuable by the management that the hard costs.

Define Phase

Project Charter

Sponsor: VP of Sales
Black Belt: John Doe
Team members:
- billing clerk (accounting)
- sales manager
- sales assistant

Scope: Find the cause of the billing errors an fix it.

Project Goal: Reduce billing errors by 90%.

Authorize Recourses: Team members will be available for up to 1 week. Conference room with media center is scheduled for this time.

Stakeholders: Sales organization, accounting, customers.

Process Map

The team met and mapped out the billing process based on the knowledge of the team members.

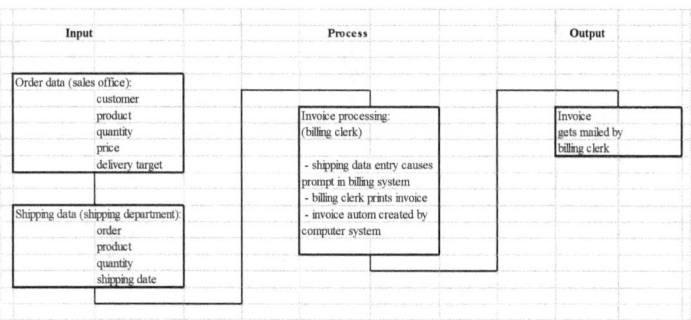

Input	Process	Output
Order data (sales office):	Invoice processing:	Invoice
customer	(billing clerk)	gets mailed by
product		billing clerk
quantity	- shipping data entry causes	
price	prompt in billing system	
delivery target	- billing clerk prints invoice	
	- invoice autom created by	
Shipping data (shipping department):	computer system	
order		
product		
quantity		
shipping date		

Measure Phase

Exploratory Data Analysis

To get the picture of the status quo the data for the last year was pulled by month and plotted the data, using the percentage of invoices in error because of seasonal changes, which are eliminated by using percentages. Getting the data was no problem since corrected invoices keep their invoice number and get a C attached to it. In the data section Number of Errors was used, which really means Number of invoices in Error, neglecting the possibility of multiple errors on one invoice.

37

	Number of				Error %	Median	UCL	LCL
	Invoices	Errors						
Jan	521	31	January		5.95%	5.88%	7.44%	4.32%
Feb	496	29	February		5.85%	5.88%	7.44%	4.32%
March	563	33	March		5.86%	5.88%	7.44%	4.32%
April	511	28	April		5.48%	5.88%	7.44%	4.32%
May	507	36	May		7.10%	5.88%	7.44%	4.32%
June	475	28	June		5.89%	5.88%	7.44%	4.32%
July	465	25	July		5.38%	5.88%	7.44%	4.32%
August	480	26	August		5.42%	5.88%	7.44%	4.32%
September	565	32	September		5.66%	5.88%	7.44%	4.32%
October	535	36	October		6.73%	5.88%	7.44%	4.32%
November	542	34	November		6.27%	5.88%	7.44%	4.32%
December	530	32	December		6.04%	5.88%	7.44%	4.32%

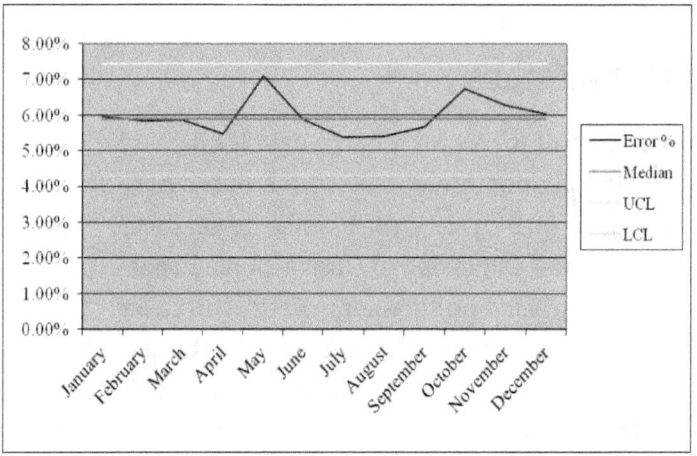

The data shows a rate of 6% invoices in error, what is indeed not acceptable.

The plotted data also shows that the process is stable / in statistical control - since it stays within the control limits of +/- 3 standard derivations and shows no trend.

No special causes seem to be present; since all values are within the control limits and no real extremes are present.

Pareto Analysis

Looking into the cause of invoices in error, the last 2 month of invoices in error were analyzed. The analysis of these 66 invoices shows clearly that the majority of errors (over 90%) come from pricing errors.

Error causes

	Price	Quantity	Other	Total
count	61	4	1	66
percentage	92.42%	6.06%	1.52%	100.00%
cumulative	92.42%	98.48%	100.00%	

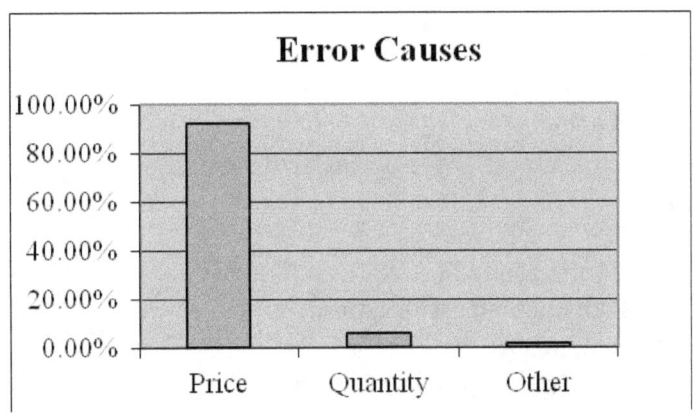

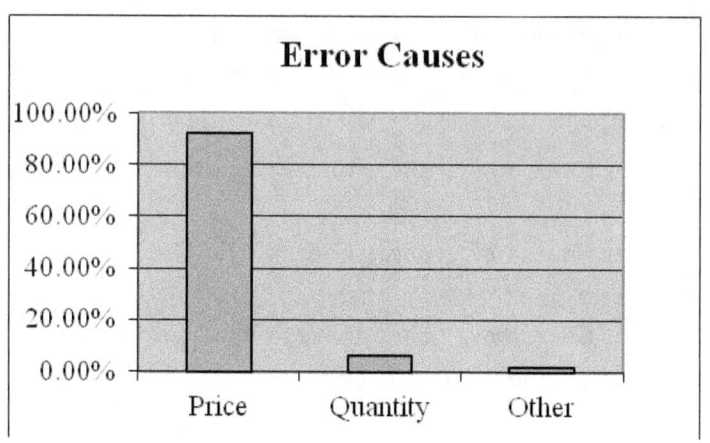

Presenting the data and these findings to the team, all agree to limit all further action to the pricing errors.

Analyze Phase

Brainstorming

The team used a brainstorming session to generate a first list of possible causes of the pricing errors. The results are:

- wrong order used
- prices not updated in the system
- order data wrong
- computer system calculates wrong

These finding seem not to be sufficient so they will be systemized in a Cause and Effect (Fishbone) diagram - also called Ishikawa diagram. This way the team was forced to more structuralized thinking and was able to come up with more possible causes and elaborate more deeply into underplaying causes.

Cause and Effect / Fishbone Diagram

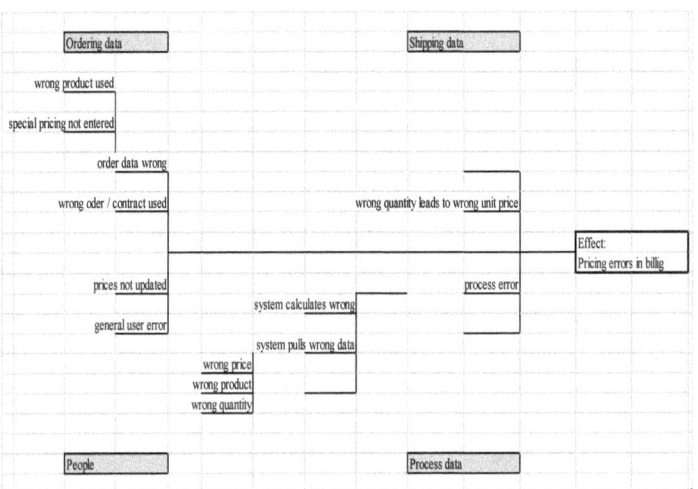

The Cause and Effect Diagram has brought up more possible causes, but didn't help in identifying the problem. Therefore it was agreed to go into a deeper analysis by going through each of the 61 invoices identified in the first Pareto Analysis to identify the exact causes of each error. This involved in many cases going back to the person involved in the invoice correction. The results shown point clearly to a computer system error since the system pulled the wrong prices. It is determent that the system does not use the discounted prices, which are given to customers based on their order quantity, if the shipment contains only a part of the order. This is clearly a system design flaw.

Causes of Pricing Error			
	A	**B**	**Total**
count	60	1	61
percentage	98.36%	1.64%	100.00%
cumulative	98.36%	100.00%	
A: system pulls wrong price			
B: special pricing not entered			

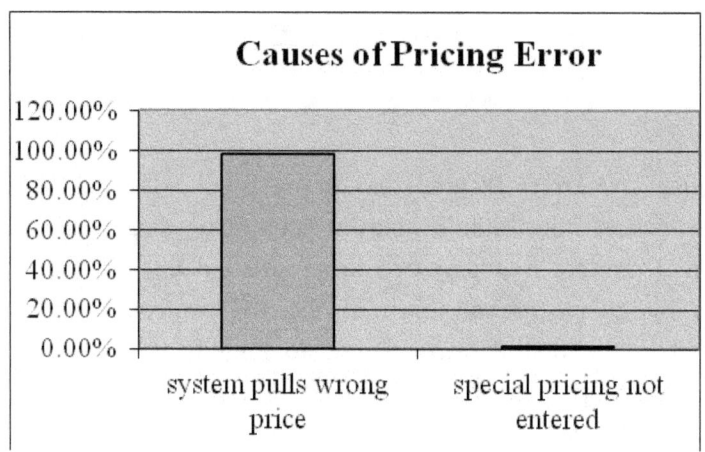

The Team reported these findings to the sponsor who

agrees to continue based on the findings.

Improve Phase

With this knowledge the team leader consulted with the IT department which checked the logic behind the invoice price. They confirm that that the quantity discount is taken from the actual delivery and not from the order. The obvious solution, to change this would be easy.

Sponsor got informed and signed off on the proposed software change, Director of IT concurred.

The logic change request was given highest priority by the Director of IT. Therefore the logic in the software was changes within the day.

The risk of implementation was low since the software change was minimal and only two tables were involved. The programmer tested the program hand had no objections do go life.

Since the implementation involved only "behind the scene" processes, there is no need to rewrite procedures or retrain users.

For the first day of been life, we analyzed three 'regular" invoices, the ones were order quantity equals shipment quantity, and all invoices there the order quantity is different from the shipped quantity to make verify that the problem got fixed and no new problem occurred due to the software changes.

Control Phase

Based on the invoice error rate from the *initial* Data Analysis of about 6% and from the Pareto Analysis of 92% been pricing errors due to the problem fixed, we estimated that about 336 invoices which normally would have been in error will be correct from now on, resulting in annual savings of $33,600. For sure this will have to be verified in the future.

At the last meeting of the Six Sigma project team was used to update the documentation of the project, presentation to the sponsor and to schedule a one hour follow meeting for about three month later (early April) to verify the savings. The black belt volunteered to collect the data for the verification of the results and to present them at the follow up meeting.

Since the project happened in January, team leader decided to let the process run for the rest of the month and use February and March data.

The count was 503 invoices with 0 in error for February and 512 invoices with 2 in error for March, what confirmed that the project was effective and the savings above are realized. At the follow up meeting it was decided that the project was a full success and that there is no need for further intervention.

44

Book Two

Lean

PREFACE

When I was getting my Six Sigma Black Belt, there was the relatively new thing around called Lean Six Sigma. Trying to get more information on it, I found that there was a lot of literature out there where it seemed that they took a regular book on Six Sigma and just replaces the phrase Six Sigma with Lean Six Sigma. I took me some time to realize that Lean Six Sigma is not the correct terminology at all. There is Six Sigma, there is Lean, and there is Designed for Six Sigma. They are all different approaches but they complement each other very well. They are all improvement tools, but that's all they have in common. This statement will set me aside from the mainstream literature, but I don't worry about this.

Six Sigma is a rigorous process comprised of DMAIC (Define, Measure, Analyse, Improve, Control), designed to reduce the errors (mistakes) to a rate of 3.4 per million opportunities or better. That's all it is about. **It's the war against variation**. It seems that there are more books written on Six Sigma (Lean Six Sigma) than there are stars in the universe, so I won't elaborate on this any further. If you want to read about Six Sigma, I recommend Thomas Pyzdek: The Six Sigma Handbook, or for the ones who just want a short overview Jay

Arthur: Lean Six Sigma DeMystified. For an even easier entry into the subject there is Barbara Wheat/Chuck Mills/Mike Carell: Leaning Into Six Sigma.

Lean on the other hand is a way of thought and a set of tools to eliminate waist (MUDA). **It's the war against waist**. Waist is defined as waiting time, extensive storage, unnecessary processes or process steps, e.g.. It is to eliminate everything what does not create value for the customer and therefore the customer is not willing to pay the extra cost for. Just-in-time delivery / supply chain management is one of the results of this thinking. The most common tools of lean are 5S (Sort, Set in order, Shine, Standardized cleanup, Sustain), Constraint Management, and the Pull System.

My point is not to favor one of the approaches above over the other one; each one has its value. It's the art of selecting the right set of tools for each problem. A hammer is a great tool, but sometimes a screwdriver or a wrench will do a better job.

LEAN

As mentioned before, Lean is the war against waist (MUDA). The goal is to reduce (non value added) activity and inventory. Therefore the focus is on
- organization of the workplace
- optimization (reduction) of inventory
- value stream
- optimization of the process steps

and at the same time it thrives for high customer satisfaction (of internal and external customers) as well as the reduction of lead time, CLT (Customer Lead Time) and PLT (Process Lead Time).
- CLT is the time a product or service needs from order to delivery (in the eye of the customer).
- PLT is the time a product needs for been produced for start to finish.

It is a change in the mindset form "If you build it they will come" (mass production) to "When they come, build it fast" (Lean production). This will take care (eliminate) unnecessary inventory (waist).

Waist does not create any value in the eye of the customer; the customer is not willing to pay it. An organization with a lot of waist will ultimately fail because there will be others with less waist and the customers will prefer them due to lower prices and equal or better quality of their output. Waist also introduces more errors into the system.

The 7 areas of waist are defined as TIMWOOD:
- Transport
- Inventory
- Movement
- Wait
- Over processing
- Over production
- Defects

The most common symptoms of waist are
- It takes multiple resources to answer customer inquiries
- Recurring problems and fixes
- No short term improvements
- Wrong inventory at the wrong time
- Problems / quarrels between departments
- Intuitive decisions instead of fact based decisions
- MUDA (the problem) is often moved on to the next department

Each organization has value added and non-value added processes:

- CAV (Customer Added Value) like assembly, mining, cutting. The customer is willing to pay for this process.
- BNVA (Business Non-Value Added) like sales, order taking, accounting, internal controlling. The customer is not willing to pay for these, but they are necessary in order to create CAVs.
- NVA (Non-Value Added) like repair of non-conforming products, extra transportation, waiting time, excessive inventory, or rework. The customer is not willing to pay for these and they are not necessary for the business process. They are waist.

It is essential to verify that an improvement adds value to the customer (Voice of the Customer). Pyzdek describes on page 706 of The Six Sigma Handbook how a project team eliminating weld dents in the production process of supermarket shelves is just to learn that the customers don't care about weld dents at all. The customers were not even aware of weld dents. The customers concern was something completely different nobody at the manufacturer cared about before.

The purpose of Lean is to reduce this waist. In order to calculate the efficiency of a process, the PCE (Process Cycle Efficiency) we divide the CAV by the Cycle time and multiply this by 100. As close this percentage is to 100%, as better the process is. This gives a good guideline on which process to focus first (most potential).

One key element of Lean is to first go after the "low hanging fruits". This usually doesn't require a sophisticated toolset and gives early victories, which is important for the morale of the team and the support from management.

Especially for the support from management it is essential to document all improvements. It is helpful to create a standardized from for it. It should contain the following categories:
1. Before – the situation found before improvement
2. Error – what was wrong with the situation
3. Activity – what was done to improve the situation
4. Responsible – who is responsible for the work area (supervisor), the date / date range of the improvement
5. After – the situation after the improvement.

Sometimes simple photos will be sufficient to document the before and after.

The following will introduce a set of tools which are helpful to become a lean organization. As mentioned before, which tools to use depends always on the individual situation.

PROCESS MAPPING

It is essential to have the processes of the organization documented and their interaction mapped out. Below is a simple example of a process map for making breakfast. Please note that the details are only outlined for one activity at each level. Each activity should be mapped in detail.

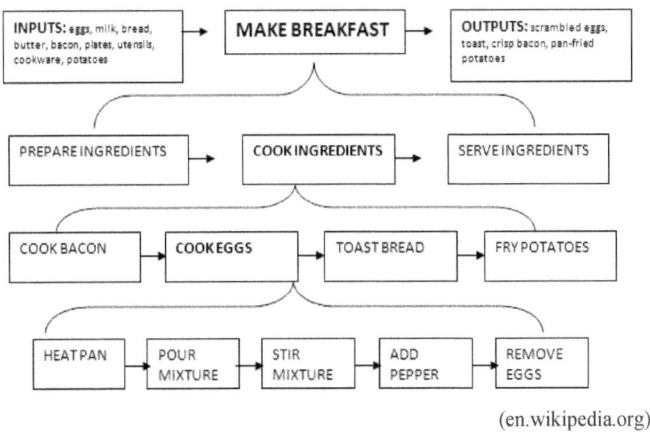

(en.wikipedia.org)

The basis of every evaluation of a process is looking at the process mapping, comparing if it still reflects reality, and then evaluate if all the steps shown are efficient and effective.

I have seen plenty of sub processes and process steps which were obsolete, didn't produce any value, and were only preformed because it was done this way for ages or just because it was included in the job description, handbook, or process map.

Process Mapping is primarily targeting the waist form of over processing.

VALUE

The first thing to do when considering a Lean approach this defining what activity creates value and which is not. Value in the eye of the customer! In the Lean approach only what the customer wants and wants to pay for has value. Everything that does not have value for the customer does not create value and therefore is waist. I order to have a clear picture of value it is important to solicit the input of the customer. This can be done with interviews or questioners. Just assuming what the customer wants and wants to pay for can be disastrous. Therefore it is essential to spend some time with the customer in order to understand the customer's perception of the products and services as well as the customer's needs.

VALUE STREAM MAPPING

Value Stream Mapping is similar to the Process Mapping; just the view is now from the perspective of the customer (internal or external customer). Is CAV produced? BNVA or NVA? The customer defines value, not the department creating it. Since most businesses have grouped work into functional silos, the temptation is great that every department defines the created value as they please. This is wrong.

The role of the value stream mapping is
- define value from the customer's view
- map the current state of the value stream
- apply the tools of Lean in order to eliminate muda (waist)
- map the future (desired) state process
- develop the transition plan
- implement the transition plan
- validate the new process

To map the value stream
6. Start by indentifying customer needs and end with satisfying them.
7. Use square Post-it notes to layout the processes

8. Use arrow Post-it notes to show delays
9. Place activities in the correct order
10. Identify inventory levels carried between each step.

Depending on the complexity of the operation, the value stream will be mapped for each process.

Current-State Value-Stream Map

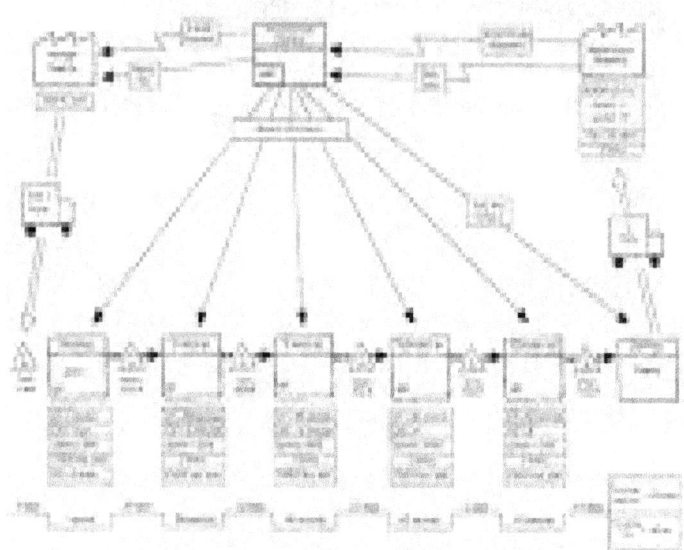

Future-State Value-Stream Map

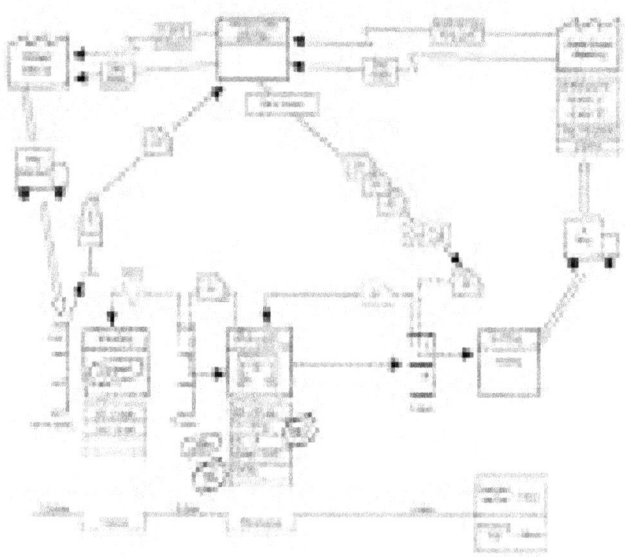

58

As shown above, the elements of a complete value stream map are:
- Process diagram
- Material flow
- Information flow
- Process data
- Process Lead Time and Value Add Time

One byproduct of value stream mapping as well as process mapping is that it has the potential to show how much time a product is sitting around waiting for the next step. This becomes obvious when one follows the product from start to finish. Often this is an eye opener.

Value Stream for a Project with Remote Testers and Customers: How long does it take to finish a feature?

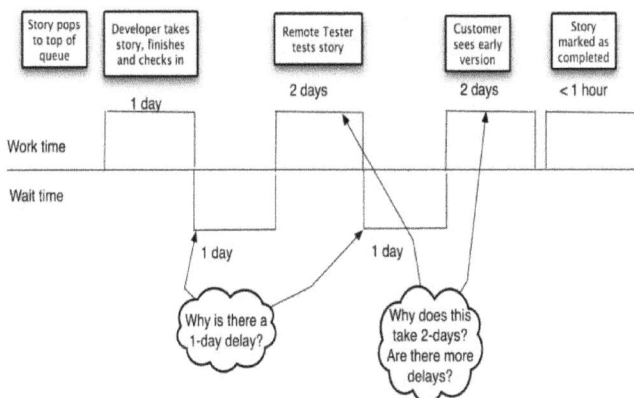

From the process mapping and value stream mapping it's just a small step to a **Value Added Flow Analysis**. Across the top row the persons or department involved will be listed, along the left hand side the major steps in the process (planning, doing, checking, and acting to improve). Starting from flow charting the process, every action, decision and arrow in the flow chart will be evaluated to identify the non-value added activities.

JUST IN TIME (JIT)

One of the key elements of reducing waste is JIT. This will reduce the inventory, one mayor source of waist by having all components arrive just in time when needed, eliminating the need for inventory, storage, and everything else what comes with it.

JIT is in essence the planning of delivery on the precise moment the product is needed. The downside is of cause that if there is a problem with the delivery, production will be stopped.

There are multiple books written on JIT and supply chain management, so I will not elaborate here further on this matter.

FOLLOW THE PRODUCT

Physically following the product is a great tool to identify waste. In this case waste of unnecessary movements and the waste of wait time.

Just following a product, an actual product or a form to be administered or approved from start to finish through the organization can be a real eye-opener. Documenting each step with actual times is a great way to update ones process mapping.

5S

The 5S metrology is one of the most used programs within the Lean concepts. It is universal, which means it can be successfully used in almost every situation. 5S is at first used to organize the workspaces, but can also be expanded into other areas. It stands for:

1. Set in order – establish an order
2. Sort – check what's there
3. Shine – clean and inspect
4. Standardize – set new standards
5. Sustain – look for improvements

This set of tasks is usually done in 3 phases; with Phase 1 containing set in order, sort, and shine, Phase 2 with standardization, and Phase 3 with sustaining.

5S - Set in order

- Clean up all unnecessary things.
- Don't store anything on the ground.
- Keep walkways free of stuff and debris.
- Don't put things against walls, machines, e.g.
- Keep the work surface free of clutter.

5S - Sort

- Keep frequently needed items in arms reach.
- Have a designated place for each tool and materials (mark that space accordingly so one can see if it is missing) – this prevents the need

of digging for a long time in the tool chest in order to find a specific tool.

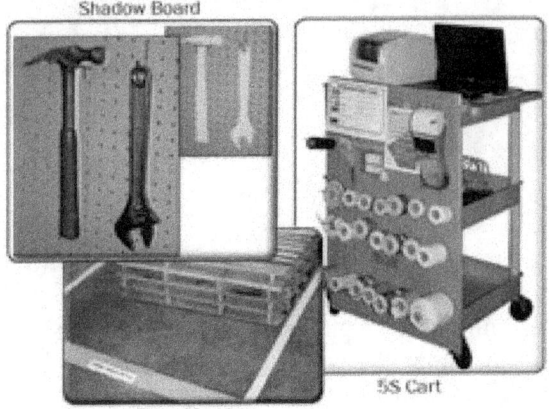

Shadow Board

Floor Marking

5S Cart

www.duralabel.com

- Keep an inventory of drawers, tool chests, storage spaces, e.g.
- Store shared tools and supplies in a central space,

It is important that every item is retrievable in a minute or less.

5S - Shine
- Clean up dirt, debris, and trash around the work space and machines.
- Keep all tools and machines clean and in good working order.

Cleaning enables the detection of weaknesses on machines and in the working environment in an early stage. The benefits are the avoidance of
- accidents
- unexpected breakdowns

- hidden expenses due to inefficient operation of machines (higher consumption of energy or other recourses)
- lower quality due to higher variance

because repair or maintenance can be scheduled before serious consequences arise. For example the early detection of a leak of a bearing or seal in a hydraulic system and appropriate action (repair request) will prevent a breakdown or accident.

5S - Standardize
- Set up of cleaning and maintenance schedules and keep documentation of it visible.
- Identify the persons responsible for each task / area. Each team member should be responsible for his / her area.
- Define, mark, and label all storage spaces on the work desk and the work area.
- Define min and max inventory and keep this information visible.
- Create a layout in order to mark areas of responsibility

5S - Sustain
- Define the expected standard and make it visible. Create SOPs (Standard Operating Procedures).
- Continuously improve the standard.
- Make every team member responsible for compliance with the standard.
- Provide training, verify compliance, and enforce discipline.

By going thru this exercise for every work place / work space will ensure that all are well organized and ready to focus on the task, not wasting time on getting ready for basic tasks and finding the items needed.

5S is mainly targeting the waist forms of wait time and movement.

PULL SYSTEM / KANBAN

Contrary to the mostly used push system, when using the pull system the production planning is done in reverse. Starting with the output, the need of product needed is given to the preceding unit and so on. Inventory at each unit is kept to a minimum, just to get by until delivery from the preceding unit. Bottlenecks are easily identified and dealt with. The KANBAN system is a variation of the pull system in which the need is communicated by sending a card (order) to the preceding unit.

The pull system decreases the inventory and cycle time.

In order to make the pull system perfect, it includes sales. This means sales trigger production and not a budget or sales projection (market pull system).

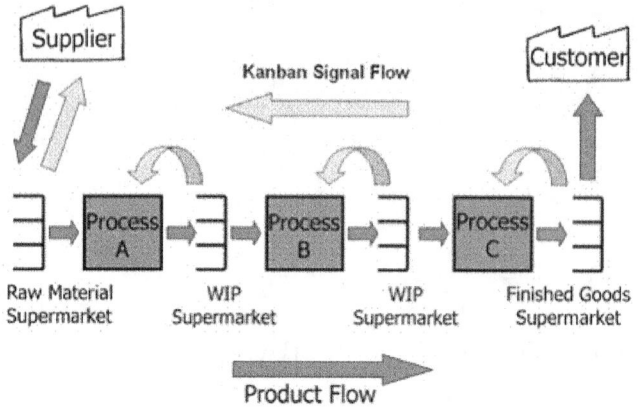

The term Replenishment Pull is used to describe a production system in which the order for production is triggered by the need to replenish a very limited inventory. This is the most likely application of the pull system.

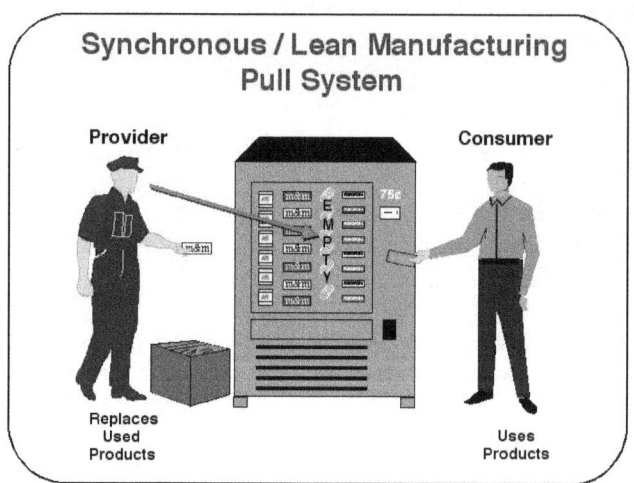

A pull system works only with a stable production process and recurring demand. When implementing a pull system the inventory will initially increase and average inventory will be reduced step by step within time. It is important to let the system work itself and not to manual interfere with it. Manual interference will destroy the system and create a chaos. When implementing a pull system, it is good practice to start with small and simple areas to gain expertise and trust in the system before a plant wide rollout.

The pull system is primarily targeting the waist form of overproduction and inventory.

VISUAL CONTROL MECHANISMS

Visual control mechanisms provide a visual overview of the status operation, department, or work area. Their purpose is to give relevant information in a quick and easy manner. It starts with the sign of days without accident and ends with the control board of mission control. There is no limit to it, as long it is relevant and not overkill. Too much information is rather confusing than helpful. Visual control mechanisms should be kept simple.

If it isn't measured it will not happen. It's like in quantum physics; a particle is a wave and at the same time matter, the final decision if wave or matter will be made when it's observed (Heisenberg's uncertainty principal). Therefore what is important for the organization / operation will have to be measured in order to keep it in the mind of everybody. Sales will keep track of overall sales or sales by representative, a non-profit will track the level of donations, the production department will track production figures, yield, e.g..

A friend of mine once implemented a control system containing of red, orange, and green lights above every machine on a plant floor. Green was for working fine, red was for not working and orange was for needing supervisor attention / help. This way any floor supervisor could see right away how the plant was performing and where attention was needed.

KAIZEN

KAIZEN (a registered trademark of KAIZEN Institute, Inc.) means continuing improvement. This is a philosophy as well as a process. Since everything in the world is evolving, everything has the potential of improvement. The business environment is constantly changing, new technology becomes available and prices change.

KAISEN can be practiced aggressively by periodically going thru the SOPs (Standard Operating Procedures) and looking for ways to improve them (mainly by experimenting) or more passively by encouraging the process owner to think of ways to improve. The steps are:
1. Someone has an idea for doing the job better (plan).
2. The impact of the idea will be investigated thru experiments and tests (do).
3. The results of the experiments will be evaluated in order to verify that the idea brings an improvement (check).
4. If the idea offers an improvement the SOP will be changed accordingly (act).

In general KAIZEN will be implemented and executed by quality improvement teams as each project should be of a short and per determent duration.

MANAGEMENT OF CONSTRAINTS - BOTTLENECKS

Each organization faces a multitude of constrains. There are legal, economic and physical constraints. This all is part the TOC (Theory of Constraints). It is like with a chain, there will be a weak link, which needs to be identified and strengthen, and the chain is improved. Then the whole chain will be re-evaluated. In general, there are Five Focusing Steps in order to deal with constraints:

1. Identify (the systems constraint)

2. Exploit (how to get the most out of the systems constraint)

3. Subordinate (subordinate everything under the exploit decision)
4. Elevate (if the system constraint still exists, increase capacity)

5. Go back to step 1.

There are two constraints an organization should deal primarily with:
1. Customer demand
2. Bottlenecks

Customer demand

In order to create customer demand it is essential to listen to the VOC (Voice Of the Customer). As mentioned before, it can be gathered by surveys, customer interviews, or reports from the sales organization.

There is one challenge with the data collection for the VOC: Biased data collection. Bias happens when the data collection is tampered with in order to get a desired data sample. I remember back in Business School when a friend was doing a marketing project and called me to come by to fill out a survey because he needed some specific profile and answers for his analysis. In order to be reliable the data cannot contain bias and it is important to collect all data to the same standard.

Bottleneck

A potential bottleneck is easily identified in the process or value stream mapping as well as just going on the floor and looking were inventory piles up and were insufficient supply slows down the process.

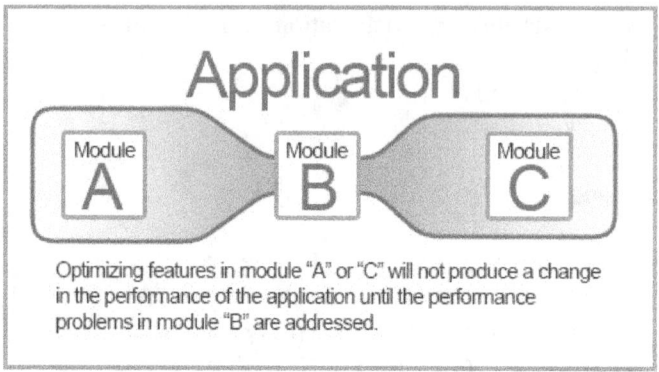

Optimizing features in module "A" or "C" will not produce a change in the performance of the application until the performance problems in module "B" are addressed.

But due diligence requires to verify that it is a real bottleneck before taking action. One has to understand the whole process in order to fix it.

There may be a bottleneck on the paper in the process or value stream mapping, but if the facility never or rarely reaches the capacity of the bottleneck, it makes no sense to act on it at the present time.

If production slows down due to leak of supply or inventory piles up at one station, there may be another reason for it too. A short seasonal peak or the unplanned downtime of a machine may be responsible.

It is good practice to find out the root cause of the problem before acting to solve it. We want to cure the cause, not the symptom.

SMED

SMED (Single Minute of Exchange Die) focuses on minimizing the time it takes to change a machine from one (product) application to another one. This becomes more and more a critical issue as the global environment requires more flexibility from every player.

Time is money and every minute a machine is down for set up and changes costs money. Therefore it is worth spending some time to minimize this downtime. This is the arena where the best will show off and succeed and the rest will lowly disappear from the market.

Just as an example: How long does it take to change 4 tires, clean the windshield, and fill up the car? Half an hour? One Hour? – Well, in the Formula I racing business it's about 7 seconds. If it takes more than 10 seconds, people will get fired. There we see the difference between the professionals and the want to be professionals. "But on this machine this doesn't work this way, this is completely different, it is not possible…" – what a lame excuse. It is possible! One just needs to want it bad enough to happen and it will be possible.

Let's analyze the Formula I example. This speed doesn't come out of the blue.

1) They planned for it – every detail, every move of the whole team. Like a ballet.
2) They prepared for it – they have every item they will need on its designated spot. No need to look around for tools e.g.
3) They practiced for it – practice, practice, practice; like a military drill. No need to think about anything, every movement is an automatic reaction.
4) They video tape it every time, analyze the video and look for waist to eliminate and areas to improve. It's constant improvement (Kaizen)

This can be done for each process, even changing or setting up a machine. Minimizing the time needed benefits the flexibility as well as the overall cost situation.

Setting a goal of time reduction is essential. But it is important to keep it in realistic perspective. A good start is aiming for a 25% reduction within 3 months. Giving the resources for the steps above will enable the team to accomplish this. If this is accomplished, another aim for additional 25% to 30% should follow.

There is no general rule on how fast one should be, it always depends on the organization, its cost situation, its position in the global environment, and then there will the point of diminishing returns.

REWORK – CLOSE THE "FIX IT" FACTORY

Rework: Just avoid it! Some factories established a fit-it factory in itself. This not only cost a lot of money to maintain, it lowers morale. "If I don't get it right, they will fix it for me."

It should be made the point that it is supposed to be right the first time. No excuses! This would be a good starting point to introduce Six Sigma and start a Six Sigma Project to eliminate the errors. (As I mentioned before, there are more good books written on Six Sigma Projects than one can read, so I will not elaborate on Six Sigma further at this point.)

MISTAKE PROOFING (POKA YOKE)

There are many ways to make a process or series of processes mistake proof / fool proof. The key is to reduce or eliminate the chance of human error by making it hard to ignore errors.

Color coding

One element of mistake proofing is color coding. It's already all over the place. Why not using it too? Electrical cables are color coded for a reason: blue is "hot", black is the return, and yellow/green is "ground". Red, yellow and green lights have a universal meaning and are not only been used on traffic signals. Orange vests are there to focus attention. On your faucet there is a blue symbol for cold and a red symbol for hot water. A red switch or button has its unique purpose. Everybody knows this, is there is no surprise. The same can be implemented in every other area. My pharmacy puts color coded tags on the medicine bottles, a different one for every household member so they are easily identified. Laboratories still use color codes to identify the process each sample has to go thru (even they now rely more on barcodes). Grease nipples and filler caps can be color coded to show to which maintenance interval they belong to. Handbook sections are color coded to identify different content.

Checklists

Checklists can be a great improvement tool too. Sometimes checklists are very complex, which have its legitimacy, but in this context I mean short checklists with up to 5 items. Their purpose is just to show the advancement in the process so that one can see within a second that's the status is, eliminating the need to read thru some tags or lists. Color coded tags will do the same, like triage tags do.

Design

Another way to reduce or eliminate human error is by designing components a certain way, so they cannot be mistaken and used wrongfully. Connectors come in different sizes so it is not hard to identify which ones are supposed to be connected where. An example is the different connectors and ports on a computer, the different connectors in modern motor vehicle where only one pair matched each other, the standardization of electric outlets and connectors, and many more.

Installation sets

Putting together installation sets helps to make sure nothing gets overlooked. When a set has 4 screws, one expects that all four screws will needed to be used and if one is left over, it is obvious that there still is a place for the leftover screw.

Transferring this principle to an assembly process, if the assembler uses a predetermined assembly set instead of getting all elements out of a big box, he will see right away if he missed something since some parts are left over.

ERGONOMICS

Ergonomics seems to be the step child of process improvement. It is hardly mentioned. But it is a great tool and actually a no-brainer. There are two major areas of ergonomics:

- Workplace ergonomics
- Process flow ergonomics

Workplace ergonomics

The classical workplace ergonomics deals with the ideal height of the chair in relation to the desk or workstation, the ideal placement of the computer monitor and keyboard e.g.. There is plenty of books and article written about this. OSHA in the US and the major health insurance companies in Europe usually provide free information, sometimes even on the job assessments and evaluations or online support.

But there is more to it: 'Soft factors" such as the right lighting and something I call workplace organization. It's part of the 5S element "sort". It is the organization of the workplace in a way that unnecessary tasks or sub-tasks are eliminated. Everything needed should be placed within easy reach. This avoids the unnecessary reaching, bending, looking, searching, and sorting.

Process flow ergonomics

Almost every process has preceding and downstream processes. They all should be in harmony, going hand-in-hand, without delays and long ways to move from one step to the next. They should – but do they?

I remember from my hospital visits when I was young that the ER and Radiology were on the opposite site of the building, preferably even on separate floors. Perfect, since most accident victims will have to get an X-Ray. That was in the old days. Luckily they now improved even further than expected and have the Radiology on little carts going from patient to patient. A quantum leap improvement; now the equipment comes to the patient instead of the patient being hauled across the building, waiting in another line again.

The floor plan of the modern American house has a process flow / lean perspective too. The garage is next to the kitchen, ideal to move groceries from the car to the refrigerator in an efficient way. Ironically, most families use their garage for everything else but to park the car. But the thought is what counts.

Similar to the workplace ergonomics, process flow ergonomics avoids extra efforts by designing the processes and the lineup of their stations accordingly.

Can we learn from the above examples and redesign our processes for a better process flow?

APPENDIX A – BALANCED SCORECARD

During the 1990 the Balanced Scorecard (BSC) got introduced into the business world. The main reason for the BSC is to give management a tool to align the organization's strategy and goals with a system of metrics. It views an organization's vision, performance, and strategic goals from four perspectives:

- Financial performance

- Customer perspective

- Internal business process

- Learning and growth

Including **financial performance** perspective encourages the identification of a few relevant high-level financial measures. In particular, designers were encouraged to choose measures that helped inform the answer to the question "How do we look to shareholders?" Examples: cash flow, sales growth, operating income, return on equity.

Including the **customers' perspective** (voice of the customer) encourages the identification of measures that answer the question "How do customers see us?" - Voice of the customer (VOC.) Examples: percent of sales from new products, on time delivery, share of important customers' purchases, ranking by important customers.

Including the **internal business process** focus encourages the identification of measures that answer the question "What must we excel at?" Examples: cycle time, unit cost, yield, new product introductions.

Including the **learning and growth** perspective encourages the identification of measures that answer the question "How can we continue to improve, create value and innovate?" Examples: time to develop new generation of products, life cycle to product maturity, time to market versus competition.

Bottom line is that all important aspects an organization wants to focus on need to get measured. No measure without a scale. The more universal the scale the better it is. Everybody involved need to understand the metrics and should learn how his/her actions translate into changes on the BSC.

Half a century ago a company or process was only managed and evaluated by financial or production ratios. Nowadays that is not sufficient anymore in order to be competitive. Looking forward there will be more areas included in the BSC such as environmental concerns (carbon footprint, sustainability, e.g.), globalization aspects, and many more. It's up to one's fantasy on what to include; everything what is deemed relevant should be evaluated. But the danger is that if there are too many, one will lose focus. There is no optimum of factors to include, but if it gets over ten, it will be very likely too much and the system needs to get trimmed down to a manageable and understandable amount of factors, strategies, and goals. Focus on the significant is the key for success.

Even as the BSC system should be flexible, its value increases when comparing the score to historical data (how did we improve) or industry data (how are we performing next to our peers). The key is, especially with industry data, that they will have to be calculated exactly the same way.

When the BSC was introduced and became popular, it was intended to help to give guidance the executive team of the corporation. It was intended to be used on a corporate level only. But what hinters one to apply the BSC to a divisional or even departmental level? In the age of the internet, getting data is not as hard and expensive as it used to be.

As mentioned above, there are almost limitless possibilities to choose from when it comes to select which business metrics to use. Different metrics are applicable to different situations, projects, and organizations. It is most important to choose the appropriate metrics. There are some characteristics common to all good metrics:

- It measures performance over time. The metric must indicate any emerging trends, and isolated measurements at static points in a process cannot do this.

- It gives direct information about the process it quantifies. The metric must give a clear and unambiguous measurement of the process it is quantifying. This allows managers to see process status at a glance, or in the larger context of emerging trends.

- It is linked directly to an organizational performance goal. The main function of a metric is to give an indication of whether a goal is being achieved or not, so it must be linked to a specific goal to be of any use.

- It is practical. This means that the employees most familiar with the process for which the metric is being developed should be instrumental in developing the metric.

- It is easy to collect and use.

It is flexible enough to change when processes changes are improved.

So – why is the Balanced Scoreboard discussed in a book about Lean? The BSC is a strategy tool for management to align goals and strategies with metrics; making them scalable and comparable, exposing customer perception as well as strength and weaknesses of the organization. As for Lean, the goal is to focus on where the money is, eliminating waist. Thus, the BSC lays the groundwork for an effective Lean approach, directing the focus of action to the area with the most potential return.

APPENDIX B – THEORY OF CONSTRAINS (TOC)

I mentioned constrains before, on a micro level, but the "official" Theory of Constrains (TOC) addresses this problem on a more macro level using metrics instead of the more hands-on methods I showed before.

The main concept behind the Theory of Constraints (TOC) is that all systems have at least one element that limits them. This restrictive element, a constraint, prevents a process from producing infinite output and stops it from being any better than they currently are. As mentioned before, a good illustration is to compare processes / systems to chains. A chain has a weakest link that prevents it from being better or stronger than it currently is. A system or process also has a weakest link limiting it. The remaining non-limiting system elements are compared to the additional links in a chain. As there can be only one weakest link or constraint limiting at a time, the remaining links, or system elements, are termed non-constraints.

The TOC focuses on improving systems by fixing the weakest link – the system constraint – in a chain. Removing constraints may remove waste from processes. As soon as the weakest link is strengthened, it is no longer the weakest link. The next weakest link becomes the constraining factor in the system, so the process starts again, and the constraining factor must be strengthened.

Constraints prevent organizations from achieving goals. Constraints may be physical, which are easier to identify as they usually have some visible element. Limiting laws or policies are more difficult to identify, as they are often entrenched in an organization; they are accepted as fact. Identifying and remedying constraining policies results in a larger degree of system improvement when addressed.

The TOC has several basic principles:

- **Systems can be thought of as chains, with weakest links or constraints** – Systems thinking is an important part of the TOC. Strengthening any link other than the weakest one constraining the system brings no benefit. In addition, continual systems improvement is needed. This is because systems change with time, and the optimal solution needs to be updated to maintain its effectiveness.

- **Managers need to understand a system before changing it** – In order to know what to change about a system, and what is the weakest link constraining it, one need a complete understanding of the whole system, its goals, and its processes.

- **The Pareto principle applies to systems** – The Pareto principle states that 80% of all problems arise from 20% of potential causes. So a few major problems are responsible for the majority of undesirable system effects.

Underlying any process are core problems. These problems cause inefficiency or waste. Key principles of the TOC regarding core problems are as follows:

- **Undesirable effects reveal core problems in the system** – The core system problems are often hidden and hard to notice. They are evident through their undesirable effects.

- **Solving core problems eliminates undesirable effects** – Fixing the undesirable effects is a short-term solution, but solving a core problem removes all of its undesirable effects. As in every other situation, it is better to treat the underlying cause (the problem) than treating the symptoms as they arise.

- **Challenge the underlying assumptions to solve core problems** – Many core problems lie hidden beneath assumptions that have come to be accepted as facts. Managers must challenge the base assumptions of the system to find real core problems.

93

The TOC involves approaching process improvement from a high-level viewpoint. Entire systems are investigated, rather than individual processes. In order to improve system performance, system bottlenecks must be reduced. Managers use three main operational measures, or metrics, to evaluate systems and their output to find where bottlenecks occur:

- **Throughput (T)** – Throughput is the rate at which a system generates money. It can be thought of as money coming in to the system.

- **Inventory (I)** – Inventory is all the money tied up in the system, either in the form of actual inventory items, or in the form of investment or reinvestment in the system and its processes.

- **Operating expenses (OE)** – Operating expenses include all the money spent on running the system and turning inventory into throughput.

Measurements are the main source of feedback for managers when deciding which changes to make. In terms of these three measurements, there are only three changes that can be made to a system – increase throughput, reduce inventory, or reduce operating expenses.

Reducing inventory and operating expenses is limited by the fact that they are both vital to the continued functioning of the system. Beyond a certain point, further reduction becomes counterproductive and begins to affect the system negatively.

Technically, throughput could be increased indefinitely. In reality, this is not true, as organizations do not have unlimited funds to support this activity, and there needs to be a market for the increased (unlimited) amount of goods produced.

Considering these aspects, it is more likely that any action taken to improve the system will focus on increasing throughput, rather than decreasing inventory or operating expenses. It is easier to increase throughput than to decrease inventory or operating expenses. The basic model for system improvement then focuses on increasing throughput, while decreasing inventory and operating expenses is a secondary aim. However, the optimal solution often incorporates elements of both of these solutions.

The following measures are often used to define process results and system output in high-level terms that are meaningful in the context of the entire organization:

- **Net profit** is the actual money that an organization makes by producing and selling a specific product. It can be calculated using this formula:

$$\text{Net profit} = \text{Throughput} - \text{Operating expenses}$$

- **Return on investment (ROI)** is a measure of how much money is being made for each dollar invested in the process. It can be calculated using this formula:

$$ROI = \frac{Throughput - Operating\ expenses}{Inventory}$$

- **Productivity** is a measure of how much is being produced by the system, in terms of saleable items. It can be calculated using this formula:

$$Productivity = \frac{Throughput}{Operating\ expenses}$$

- **Turnover** is a measure of the rate that a system turns work in progress (WIP) into saleable items, and eventual profit. It can be calculated using this formula:

$$Turnover = \frac{Throughput}{Inventory}$$

- **Throughput (T)** – Throughput is the rate at which a system generates money. It is calculated by the formula:

ABOUT THE AUTHOR

Klaus Hogreve was raised and educated in Germany and is currently living in Southern California.
He graduated from the University of Lüneburg, Germany, and holds the CMA, CFM well as the Six Sigma Black Belt certifications.